THE MALLEE IN FLOWER

I. R. McCANN

For Thelma

PREFACE

"There is a pressing urgency to conserve the slender tracts of remaining native Mallee bush, those shrinking islands in a sea of European land uses that retain what is left of the original heartlands and their species."

Julian Cribb, "The Forgotten Wilderness"
The Weekend Australian, May 27-28, 1989

The Victorian National Parks Association was formed in 1952. One of its objectives is "to promote the education of the community in matters relating to the conservation of indigenous flora and fauna and other natural resources". It is in this context that we publish this important work by the naturalist, Ian McCann.

Two-thirds of the Mallee country of south-eastern Australia has been cleared for agriculture. Of the original 1000 plant species, 200 have been lost or are now endangered. This extensive loss of native vegetation has resulted in problems of salinity and wind erosion, which are affecting the landscape, farm productivity and land values.

It is not too late to halt this degradation. Victoria already has many national and other parks which protect extensive tracts of the Mallee. Substantial additions to this reserve system have been recommended by the Land Conservation Council (LCC) in its Mallee Area Review Final Recommendations of August 1988. The VNPA hopes that these recommendations will ultimately be accepted by the Government.

However, parks and reserves, by themselves, are not enough. The retention of native vegetation on farms and roadsides also plays a key role in the fight against land degradation. We pay a tribute to the work already being done by many private landholders and the Department of Conservation, Forests and Lands in its Saltwatch, Tree Victoria, Landcare and other community programmes. Managers of both public and private land are now seeing compelling reasons to save remnants of existing Mallee flora, and to revegetate degraded areas using indigenous species.

We thank Ian McCann for his many years of painstaking work in photographing more than 400 species and we hope that this book, by stimulating local interest in the flora of the Mallee, will be a catalyst for a change in attitudes towards the Mallee. The 'useless scrub' will become 'valuable bushland'.

By joining the Victorian National Parks Association you can help save the Mallee and other areas under threat. Membership and other enquiries may be made to:

The Victorian National Parks Association
247 Flinders Lane
Melbourne 3000. Tel (03) 654 6843.

Joan Lindros
PRESIDENT

VICTORIAN NATIONAL PARKS ASSOCIATION Inc.

INTRODUCTION

The author's aim has been to provide a pictorial introduction to the plants of north-western Victoria for use by people living in this interesting region and by the growing volume of people who visit the area for recreation.

All of the plants depicted grow in this area and in the mallee country of eastern South Australia and south-western New South Wales. An attempt has been made by means of a letter code to indicate the distribution of each plant by noting the main reserves in which it has been recorded. These reserves are shown on the map on the inside of the rear cover and are coded as follows:

W Wyperfeld National Park
B Big Desert Wilderness
P Pink Lakes State Park
H Hattah-Kulkyne National Park
M Mungo National Park

It is certain that the available lists of plants in these reserves are incomplete and particularly so in relation to the Big Desert Wilderness and Mungo National Park. Plants growing in the Big Desert Wilderness are likely also to be found in the contiguous Scorpion Springs, Mount Shaugh and Ngarkat Conservation Parks in South Australia.

The common names used have been taken from A. C. Beauglehole's "Victorian Vascular Plants Checklists", 1980, and the scientific names from "A Census of the Vascular Plants of Victoria" by S. J. Forbes and J. H. Ross, Second edition 1988.

Most of the illustrations are reproduced approximately natural size. As a further aid to identification a brief description, height range and the months or season of its main flowering period are given for each of the plants illustrated.

A number of established alien plants have been illustrated and are indicated by an asterisk.

Significant changes are taking place in the species composition of our Mallee vegetation. Rabbits have long been established and sheep have been run over much of the public land. Their selective grazing and browsing and the breaking of the delicate soil crust have resulted in many native plants being reduced in number and in the establishment of alien species. The watering points provided for sheep have also served the rabbits, feral goats, feral bees and kangaroos. They in turn have influenced the native flora and the original fauna. Plants that have been classified as vulnerable, endangered or rare in Victoria have been indicated by obelisks; † = vulnerable, †† = endangered, ††† = rare.

Yet the Mallee vegetation is still diverse, interesting and, responding to the season, can be very beautiful. Each flower has its own particular charm and better acquaintance will enhance the pleasure it provides.

†††Woolly Clock-fern
Cheilanthes lasiophylla
Adiantaceae.

Fronds erect to 15cm. -B---.

Pacific Azolla
Azolla filiculoides
Azollaceae.

Free-floating aquatic. W--H-.

Common Nardoo
Marsilea drummondii
Marsileaceae.

Aquatic. Fronds to 30cm. W--H-.

Short-fruit Nardoo *Marsilea hirsuta* Marsileaceae. Aquatic. Fronds to 15cm. W--H-.

Austral Pillwort *Pilularia novae-hollandiae* Marsileaceae. Fronds grass-like to 6cm. Damp ground. W----.

†**Adder's Tongue** *Ophioglossum polyphyllum* Ophioglossaceae. Fronds to 9cm. ---HM.

White Cypress Pine *Callitris glaucophylla* Cupressaceae. Tree to 13m. ---HM.

Slender Cypress Pine
Callitris preissi
Cupressaceae.

Tree to 17m. W-PH-.

Scrub Cypress Pine
Callitris verrucosa
Cupressaceae.

Shrub or small tree to 6m. WBPHM.

Swamp Lily
Ottelia ovalifolia
Hydrocharitaceae.

Aquatic perennial. Summer. ---H-.

***Onion Weed** *Asphodelus fistulosus* Liliaceae. Annual. Invades disturbed ground. 20-60cm. Aug.-Dec. ---H-.

Leek Lily *Bulbine semibarbata* Liliaceae. Perennial. 15-30cm. Spring. W-PH-

†††Garland Lily *Calostemma purpureum* Liliaceae. Perennial. 20-50cm. Jan.-April. ---H-.

Nodding Chocolate-lily *Dichopogon fimbriatus* Liliaceae. Slender perennial. 30-50cm. Dec. W----.

Yellow Star *Hypoxis vaginata* Liliaceae. Perennial. 10-20cm. Aug.-Oct. W-PH-.

Mallee Fringe-lily *Thysanotus baueri* Liliaceae. Perennial. 15-30cm. Oct.-Dec. W-PH-.

Branching Fringe-lily *Thysanotus juncifolius* Liliaceae. Perennial. 25-50cm. Nov.-Jan. WBP--.

Twining Fringe-lily *Thysanotus patersonii* Liliaceae. Climbing perennial to 90cm. Sept.-Nov. WB---.

Yellow Rush-lily
Tricoryne elatior
Liliaceae.

Wiry perennial to 40cm. Oct.-Feb. W--H-.

Early Nancy
Wurmbea dioica
Liliaceae.

Perennial. 5-20cm. Aug.-Sept. W-PH-.

Pink Fingers
Caladenia catenata
Orchidaceae.

Perennial. 7-25cm. Spring WB---.

Green-comb Spider-orchid *Caladenia dilatata* Orchidaceae. Perennial. 15-40cm. Spring WBPH-.

Daddy Long-legs *Caladenia filamentosa* Orchididaceae. Perennial. 15-30cm. Sept.-Oct. W----.

Purplish Beard-orchid *Calochilus robertsonii* Orchidaceae. Perennial. 20-45cm. Oct.-Nov. W----.

Horned Orchid *Orthoceras strictum* Orchidaceae. Perennial. 15-45cm. Nov.-Feb. -B---.

Tall Leek-orchid *Prasophyllum elatum* Orchidaceae. Perennial. 30-90cm. Oct.-Nov. W----.

†††**Tawny Leek-orchid** *Prasophyllum fuscum* Orchidaceae. Perennial. 15-30cm. Oct.-Nov. W----.

Mallee Midge-orchid *Prasophyllum nigricans* Orchidaceae. Perennial. 10-25cm. Mar.-May WB-H-.

Rusty-hood *Pterostylis biseta* Orchidaceae. Perennial. 10-35cm. Oct.-Dec. WB-H-.

'Long-tongued' Greenhood
Pterostylis dolichochila
Orchidaceae.

Perennial. 10-20cm. June W----.

Dwarf Greenhood
Pterostylis nana
Orchidaceae.

Perennial. 5-10cm. July-Oct. ---H-.

Bearded Greenhood
Pterostylis plumosa
Orchidaceae.

Perennial. 15-30cm. Sept.-Oct. -B---.

Midget Greenhood
Pterostylis mutica
Orchidaceae.

Perennial. 10-15cm. July-Sept. WBPH-.

Rabbit-ears
Thelymitra antennifera
Orchidaceae.

Perennial. 12-20cm. Sept.-Oct. WB---.

†Azure Sun-orchid
Thelymitra canaliculata
Orchidaceae.

Perennial. 15-30cm. Oct.-Nov. W----.

Scented Sun-orchid *Thelymitra nuda*
Orchidaceae. Perennial. 30-50cm.
Sept.-Nov. WBPH-.

Porcupine Grass *Triodia irritans*
Poaceae. Prickly tussocks to 1½m.
Spring WBPHM.

Floating Pondweed *Potamogeton tricarinatus* Potamogetonaceae. Aquatic.
Summer ---H-.

Scented Mat-rush *Lomandra effusa*
Xanthorrhoeaceae. Tufted perennial to 60cm. June-Oct. WBPH-.

Desert Mat-rush
Lomandra juncea
Xanthorrhoeaceae.

Tufted perennial to 40cm. Aug.-Oct. WB-H-.

Tufted perennial to 60cm. Winter WBPH-.

Woolly Mat-rush
Lomandra leucocephala
Xanthorrhoeaceae.

Inland Pigface
Carpobrotus modestus
Aizoaceae.

Prostrate perennial. Sept.-Dec. W-PH-.

Prostrate perennial. Sept.-Nov. ---PHM.

Rounded Noon-flower
Disphyma crassifolium
Aizoaceae.

***Common Ice-plant**
Mesembryanthemum crystallinum
Aizoaceae

Prostrate annual or biennial. Oct.-Dec. --PHM.

Tangled procumbent annual or biennial. Sept.-Dec. ---H-.

***Wiry Noon-flower**
Psilocaulon tenui
Aizoaceae.

Sarcozona
Sarcozona praecox
Aizoaceae.

Procumbent perennial. Aug.-Sept. --P--.

Protrate perennial. Sept.-Jan. ---HM.

New Zealand Spinach
Tetrtagonia tetragonioides
Aizoaceae.

Lamb-tails
Ptilotus exaltatus
Amaranthaceae.

Erect perennial. 30-50cm. Spring W-PHM.

Ascending perennial. 20-30cm. Oct.-Dec. W-PH-.

Rabbit-tails
Psilotus seminudus
Amaranthaceae.

Pussy-tails
Psilotus spathulatus
Amaranthaceae.

Prostrate perennial. Sept.-Dec. W-PH-.

†Austral Doubah
Leichardtia australis
Asclepiadaceae.

Twining liane to 5m. Summer ---HM.

†Purple Pentatrope
Rhynchorrhena linearis
Asclepiadaceae.

Slender perennial climber. Summer ---H-.

Flannel Cudweed
Actinobile uliginosa
Asteraceae.

Prostrate annual. Aug.-Nov. W-PH-.

Hairy Angianthus
Angianthus tomentosus
Asteraceae.

Erect annual to 20cm. Oct.-Dec. W--H-.

Dwarf-Beauty-heads
Blennospora drummondii
Asteraceae.

Erect annual. 3-9cm. Spring W-PH-.

Variable Daisy
Brachyscome ciliaris
Asteraceae.

Annual. 5-30cm. Aug.-Dec. W-PH-.

Hard-head Daisy
Brachyscome lineariloba
Asteraceae.

Annual. 5-10cm. Aug.-Sept. W-PH-.

Rayless Daisy
Brachyscome perpusilla
Asteraceae.

Annual to 6cm. Sept.-Oct. W-PH-.

Erect or bushy perennial to 60cm. June-Oct. ---H-.

Blue Burr-daisy
Calotis cunefolia
Asteraceae.

Burr-daisy
Calotis cymbacantha
Asteraceae.

Bushy perennial to 50cm. Spring. W--H-.

Shrub to 1½m. Summer WB---.

Sticky Cassinia
Cassinia uncata
Asteraceae.

Common Sneezeweed
Centipeda cunninghamii
Asteraceae.

Aromatic perennial to 20cm. Spring-Summer --PH-.

Annual to 1m. Spring-Summer. W-PHM.

***Malta Thistle**
Centaurea melitensis
Asteraceae.

***Slender Thistle** *Carduus tenuiflorus* Asteraceae. Annual or biennial to 1m. Oct.-Jan. W--H-.

***Chicory** *Cichorium intybus* Asteraceae. Perennial to 1m. Nov.-May H .

***Saffron Thistle** *Carthamus lanatus* Asteraceae. Erect annual to 90cm. Oct.-Dec. W-PH-.

Common Billy-buttons *Craspedia glauca* Asteraceae. Annual to 50cm. July-Dec. WB---.

Soft Billy-buttons *Craspedia pleiocephala* Asteraceae. Annual to 30cm. Aug.-Nov. ---HM.

Spreading Nut-heads *Epaltes australis* Asteraceae. Prostrate annual of flooded areas. Summer ---H-.

Woolly Mantle *Eriochlamys behrii* Asteraceae. Annual 3-12cm. Nov.-Jan. ---H-.

Common Cudweed *Gnaphalium involucratum* Asteraceae. Annual to 40cm. Spring W--H-.

Scaly Haeckeria *Haeckeria pholidota* Asteraceae. Shrub to 1m. Summer WB---.

Common Everlasting *Helichrysum apiculatum* Asteraceae. Perennial. 10-50cm. Spring WBPH-.

White Everlasting *Helichrysum baxteri* Asteraceae. Erect perennial. 20-40cm. Spring WB---.

Woolly Everlasting *Helichrysum blandowskianum* Asteraceae. Erect perennial. 20-40cm. Summer W----.

Erect perennial 30-75cm. Sept.-Dec. W-PH-.

Golden Everlasting
Helichrysum bracteatum
Asteraceae.

Ridged Everlasting
Helichrysum catadromum
Asteraceae.

Erect shrub to 75cm. Nov.-Dec. W--H-.

Erect perennial. 30-50cm. Spring WBPH-.

Satin Everlasting
Helichrysum leucopsidium
Asteraceae.

Blunt Everlasting
Helichrysum obtusifolium
Asteraceae.

Erect perennial. 10-30cm. Spring WB---.

Tufted annual to 2½cm. Sept.-Oct. W-PH-.

Moss Sunray
Helipterum demissum
Asteraceae.

Grass Cushion
Isoetopsis graminifolia
Asteraceae.

Prostrate annual. Spring W-PH-.

Clustered Everlasting *Helichrysum semipapposum* Asteraceae. Erect perennial. 30-50cm. Summer WB---.

Musk Sunray *Helipterum moschatum* Asteraceae. Erect annual. 10-40cm. Spring W-PH-.

Milkwort Sunray *Helipterum polygalifolium* Asteraceae. Erect annual. 15-30cm. July-Nov. --P--.

†Fleshy Kippistia *Kippistia suaedifolia* Asteraceae. Perennial. 8-20cm. Saline soils. Oct.-Nov. --P--.

Yam-daisy
Microseris scapigera
Asteraceae.

Erect perennial. 15-30cm. Spring W-PH-.

Broad-leaf Millotia
Millotia myosotidifolia
Asteraceae.

Tufted annual. 3-15cm. Spring W--H-.

Soft Millotia
Millotia tenuifolia
Asteraceae.

Erect annual. 5-15cm. Spring --PH-.

Bushy perennial. 7-30cm. Spring W-P--.

Minnie Daisy
Minuria leptophylla
Asteraceae.

Woolly-heads
Myriocephalus rhizocephalus
Asteraceae.

Prostrate annual. Oct.-Dec. W-PH-.

Erect annual. 10-50cm. Sandy soil. Spring W-PHM.

Poached-eggs Daisy
Myriocephalus stuartii
Asteraceae.

Fringed Daisy-bush
Olearia ciliata
Asteraceae.

Semi-shrub. 18-30cm. Nov.-Jan. WBPH-.

Erect shrub ½-1m. Jan.-Mar. W----.

Woolly Daisy-Bush
Olearia lanuginosa
Asteraceae.

Mueller Daisy-bush
Olearia muelleri
Asteraceae.

Shrub to 60cm. Aug.-Oct. --PH-.

Pimelea Daisy-bush
Olearia pimeleoides
Asteraceae.

Shrub to 1m. Spring W-PH-.

Cypress Daisy-bush
Olearia teretifolia
Asteraceae.

Erect shrub to 1½m. Spring W--H-.

***Stemless Onopordon**
Onopordum acaulon
Asteraceae.

Prostrate biennial. Oct.-Dec. W-PH-.

Azure Daisy-Bush *Olearia rudis*
Asteraceae. Semi-shrub to 90cm.
Aug.-Nov. W-PH-.

Fleshy Groundsel *Othonna gregorii*
Asteraceae. Erect annual. 10-30cm.
Aug.-Oct. W-P--.

†Grey Podolepis *Podolepis canescens*
Asteraceae. Erect annual. 15-40cm.
Spring W-P--.

Wiry Podolepis *Podolepis capillaris*
Asteraceae. Branching wiry annual.
15-30cm. W-PH-.

Pleated Podolepis
Podolepis rugata
Asteraceae.

Shrubby perennial. 15-60cm. Spring ---H-.

Sticky Longheads
Podotheca angustifolia
Asteraceae.

Ascending annual. 5-20cm. Aug-Dec. WBPH-.

Stiff Pogonolepis
Pogonolepis muelleriana
Asteraceae.

Wiry ascending annual. Saline soils. 2-15cm. W-PH-.

†††**Tall Yellow-top** *Senecio magnificus* Asteraceae. Perennial semi-shrub to 1m. Winter-Spring W----.

***Sow Thistle** *Sonchus oleraceus* Asteraceae. Erect annual to 1m. Most months. W-PH-.

Common Bow-flower *Toxanthes muelleri* Asteraceae. Procumbent annual to 3cm. Aug.-Nov. W-PH-.

Woolly New Holland Daisy *Vittadinia cuneata* Asteraceae. Perennial. 10-30cm. Spring-Summer W-PHM.

Annual. 2-10cm. Sept.-Nov. W-P--.

Common Sunray
Triptilodiscus pygmaeus
Asteraceae.

†Giant New Holland Daisy
Vittadinia megacephala
Asteraceae.

Perennial. 15-30cm. Spring W----.

Erect annual. 15-40cm. Spring W--HM.

Orange Immortelle
Waitzia acuminata
Asteraceae.

Sweet Hound's-tongue *Cynoglossum suaveolens* Boraginaceae. Fragrant perennial. 15-50cm. Spring W----.

Rough Halgania *Halgania cyanea* Boraginaceae. Bushy shrub to 30cm. Spring-Summer. W--H-.

Lavender Halgania *Halgania lavandulacea* Boraginaceae. Shrub to 60cm. Spring W--H-.

***Common Heliotrope** *Heliotropium europaeum* Boraginaceae. Annual. 10-30cm. Invasive weed. Spring-Autumn W-PH-.

Erect biennial to 1m. Troublesome weed. May-Dec. W---M.

***Paterson's Curse**
Echium plantagineum
Boraginaceae.

Prostrate perennial. Lake beds. Spring-Autumn W----.

Smooth Heliotrope
Heliotropium curassavicum
Boraginaceae.

Erect annual to 30cm. Aug.-Nov. W-PH-.

Sand Cress
Pachymitus cardaminoides
Brassicaceae.

May Smocks *Harmsiodoxa blennodioides*
Brassicaceae. Erect annual. 10-30cm.
July-Oct. W--H-.

Desert Cassia *Cassia nemophila*
Caesalpinaceae. Bushy shrub to 2m.
Spring W-PH-.

Tufted Bluebell *Wahlenbergia communis*
Campanulaceae. Erect perennial.
30-60cm. Spring --PH-.

Dwarf She-oak *Allocasuarina pusilla*
Casuarinaceae. Shrub to 1m. Female
plant. Winter WB---.

Bull-oak
Allocasuarina luehmannii
Casuarinaceae.

Tree to 15m. Female plant. Summer W-PH-.

Bull-oak
Allocasuarina luehmannii
Casuarinaceae.

Tree to 15m. Male plant. Summer W-PH-.

Slaty She-oak
Allocasuarina muelleriana
Casuarinaceae.

Shrub to 4m. Spring-Summer. WB-H-.

Scrub She-oak
Allocasuarina paludosa
Casuarinaceae.

Erect shrub. 1-3m. Spring WB---.

Tree to 8m. Female plant. Summer W----.

Drooping She-oak
Allocasuarina verticillata
Casuarinaceae.

Belah
Casuarina cristata
Casuarinaceae.

Tree to 15m. Female plant. Summer --PHM.

Annual semi-shrub to 40cm. Oct.-Mar. --PHM.

Baldoo
Atriplex lindleyi
Chenopodiaceae.

Berry Saltbush
Atriplex semibaccata
Chenopodiaceae.

Prostrate perennial. Summer W--H-.

Annual. 10-30cm. Summer ---HM.

Pop Saltbush
Atriplex spongiosa
Chenopodiaceae.

Old-man Saltbush *Atriplex nummularia*
Chenopodiaceae. Bushy shrub to 2m.
Spring-Summer ---HM.

†Silver Saltbush *Atriplex rhagodioides*
Chenopodiaceae. Shrub to 1m. Oct.-Mar.
---H-.

†Kidney Saltbush *Atriplex stipitata*
Chenopodiaceae. Slender shrub to 1m.
Nov.-Mar. ---H-.

Bladder Saltbush *Atriplex vesicaria*
Chenopodiaceae. Erect shrub to 60cm.
Oct.-Mar. --PHM.

Babbagia *Babbagia acroptera* Chenopodiaceae. Perennial to 20cm. Sept.-Mar. ---HM.

Hard-head Dissocarpus *Dissocarpus paradoxus* Chenopodiaceae. Semi-shrub to 30cm. Oct.-Mar. ---HM.

Ruby Saltbush *Enchylaena tomentosa* Chenopodiaceae. Procumbent shrub. Oct.-Mar. W-PHM.

Gray Glassewort *Halosarcia halocnemoides* Chenopodiaceae. Shrub 20-90cm. Salt lakes. Spring --PH-.

†††Wiry Glasswort
Halosarcia lylei
Chenopodiaceae.

Erect shrub to 1m. Salt lakes. Spring ---H-.

Shrub to 50cm. Salt lakes. Spring ---H-.

A Glasswort
Halosarcia nitida
Chenopodiaceae.

A Glasswort
Halosarcia pergranulata
Chenopodiaceae.

Shrub to 50cm. Salt lakes. Spring --PH-.

Shrub to 1½m. Summer W-PHM.

Short-leaf Bluebrush
Maireana brevifolia
Chenopodiaceae.

Rosy Bluebush
Maireana erioclada
Chenopodiaceae.

Shrub to 60cm. Sept.-Mar. --PH-.

Shrub to 1m. Sept.-Mar. ---HM.

Shrubby Bluebush
Maireana pyramidata
Chenopodiaceae.

†††Rohrlach's Bluebush
Maireana rohrlachii
Chenopodiaceae.

Shrub to 1m. W--H-.

Hoary Bluebush
Maireana sedifolia
Chenopodiaceae.

Shrub to 1m. Summer --PHM.

Prickly Saltwort
Salsola kali
Chenopodiaceae.

Annual to 1m. Sept.-Feb. W-PHM.

Thick-head Glasswort
Sarcocornia blackiana
Chenopodiaceae.

Semi-shrub to 80cm. Salt lakes. Aug.-Nov. --P--.

Beaded Glasswort
Sarconornia quinqueflora
Chenopodiaceae.

Perennial to 50cm. Nov.-Feb. --P--.

Hairy Saltbush
Sclerochlamys brachyptera
Chenopodiaceae.

Perennial to 15cm. Sept.-Dec. ---HM.

Tangled Saltbush
Sclerolaena tricuspis
Chenopodiaceae.

Semi-shrub to 1m. Oct.-Mar. --PHM.

Pink Bindweed
Convolvulus erubescens
Convolvulaceae.

Creeping perennial. Oct.-Dec. W-PH-.

***Colocynth**
Citrullus colocynthis
Cucurbitaceae.

Prostrate perennial vine. Summer -----.

Prostrate annual vine. Dec.-April ---HM.

***Wild Melon**
Citrullus lanatus
Cucurbitaceae.

***Paddy Melon**
Cucumis myriocarpus
Cucurbitaceae.

Prostrate annual vine. Autumn. W--H-.

Prostrate annual vine. Summer ---H-.

Mallee Cucumber
Mukia micrantha
Cucurbitaceae.

Silky Guinea-flower *Hibbertia sericea*
Dilleniaceae. Shrub to 50cm. Spring
-B-H-.

Erect Guinea-flower *Hibbertia stricta*
Dilleniaceae. Shrub to 90cm. Aug.-Oct.
WB-H-.

Twiggy Guinea-flower *Hibbertia virgata*
Dilleniaceae. Slender shrub to 60cm.
July-Nov. W-PH-.

Scarlet Sundew *Drosera glanduligera*
Droseraceae. Perennial to 6cm. Spring
W----

Flame Heath *Astroloma conostephioides*
Epacridaceae. Erect shrub. 40-90cm.
May-Oct. WB---

Cranberry Heath *Astroloma humifusum*
Epacridaceae. Prostrate or erect shrub to 20cm. Autumn WB---.

Daphne Heath *Brachyloma daphnoides*
Epacridaceae. Erect shrub to 90cm.
Aug.-Oct. WB---.

Brush Heath *Brachyloma ericoides*
Epacridaceae. Bushy shrub. 30-60cm.
July-Oct. WB---.

Cleland's Beard-heath
Leucopogon clelandii
Epacridaceae.

Shrub to 50cm. Feb.-April WB---.

Heart-leaf Beard-heath
Leucopogon cordifolius
Epacridaceae.

Shrub to 1½m. Dune tops. Autumn WB-H-.

Twiggy Beard-heath
Leucopogon costatus
Epacridaceae.

Shrub to 40cm. Aug.-Nov. WB---.

Erect shrub to 1m. Summer WB-H-.

Ruddy Beard-heath
Leucopogon rufus
Epacridaceae.

Nodding Beard-heath
Leucopogon woodsii
Epacridaceae.

Erect shrub to 30cm. Winter WB---.

Shrub to 1m. Female plant. Aug.-Dec. W-PH-.

Mallee Bitter-bush
Adriana hookeri
Euphorbiaceae.

Mitchell Bertya
Bertya mitchellii
Euphorbiaceae.

Erect shrub to lm. Male plant. Aug.-Sept. WB-H-.

Pale Turpentine Bush
Beyeria lechenaultii
Euphorbiaceae.

Viscid Shrub to 1½m. Spring W-PH-.

Dark Turpentine Bush
Beyeria opaca
Euphorbiaceae.

Viscid shrub to 30cm. Male plant. Winter-Spring W-PH-.

Shrub to 90 cm. Spring WBPH-.

Common Aotus
Aotus ericoides
Fabaceae.

Leafless Bitter-pea
Daviesia brevifolia
Fabaceae.

Shrub to 90cm. Spring WB---.

Procumbent to low shrub. Spring WBPH-.

Common Eutaxia
Eutaxia microphylla
Fabaceae.

Red Parrot-pea *Dillwynia hispida* Fabaceae. Slender shrub to 30cm. Spring WB---.

Showy Parrot-pea *Dillwynia sericea* Fabaceae. Stiff shrub to 90cm. Aug.-Oct. WB---.

††**Silky Glycine** *Glycine canescens* Fabaceae. Twining perennial. July-Oct. ---H-.

Southern Liquorice *Glycyrrhiza acanthocarpa* Fabaceae. Semi-shrub. 60-90cm. Lake beds. Summer W--H-.

Running Postman *Kennedia prostrata*
Fabaceae. Prostrate shrub. Aug.-Oct.
W--H-.

††**Spreading psoralea** *Psoralea patens*
Fabaceae. Perennial to 90cm. All months
W----.

Silky Bush-pea *Pultenaea prostrata*
Fabaceae. Rigid shrub to 90cm.
Nov.-Dec. WB---.

Slender Bush-pea *Pultenaea tenuifolia*
Fabaceae. Procumbent shrub to 50cm.
Spring WB---.

Small-leaf Swainson-pea
Swainsona microphylla
Fabaceae.

Procumbent perennial. Aug.-Oct. ---H-.

†Dwarf Swainson-pea
Swainsona phacoides
Fabaceae.

Ascending or prostrate perennial. Spring ---H-.

†††Hoary Sea-heath
Frankenia crispa
Frankeniaceae.

Dense shrub to 30cm. Salt lakes. Oct.-Dec. --PH-.

Leafy Sea-heath *Frankenia foliosa*
Frankeniaceae. Dense shrub to 25cm.
Salt lakes. Summer --PH-.

Southern Sea-heath *Frankenia pauciflora*
Frankeniaceae. Shrub to 30cm. Salt lakes.
Oct.-Jan. --P--.

Small-leaf Sea-heath *Frankenia sessilis*
Frankeniaceae. Spreading shrub to 30cm.
Nov.-Jan. --P--.

Spike Centaury *Centaurium spicatum*
Gentianaceae. Erect annual. 10-45cm.
Oct.-Feb. W--H-.

***A Centaury**
Centaurium tenuiflorum
Gentianaceae.

Erect annual to 30cm. Spring-Summer W--H-.

Blue Heron's-bill
Erodium crinitum
Geraniaceae.

Annual. 8-30cm. Spring W-PHM.

Austral Crane's-bill
Geranium solanderi
Geraniaceae.

Prostrate or ascending perennial. Aug.-Dec. W----.

Shrub to 40cm. Spring ---H-.

Grooved Dampiera
Dampiera lanceolata
Goodeniaceae.

Velvet Dampiera
Dampiera marifolia
Goodeniaceae.

Shrub to 40cm. Spring WBPH-.

Shrub to 50cm. Spring W--H-.

Rosemary Dampiera
Dampiera rosmarinifolia
Goodeniaceae.

Clasping Goodenia
Goodenia amplexans
Goodeniaceae.

Semi-shrub 30-60cm. Sept.-April W----.

Perennial. 15-30cm. Aug.-Dec. WB---.

Woolly Goodenia
Goodenia robusta
Goodeniaceae.

Sticky Goodenia
Goodenia varia
Goodeniaceae.

Tangled semi-shrub to 50cm. Oct.-Dec. W--H-

Sandhill Goodenia *Goodenia sp.* Goodeniaceae. Tufted perennial to 15cm. Aug.-Jan. WBPH-.

Fairy Fan-flower *Scaevola aemula* Goodeniaceae. Procumbent perennial to 45cm. Sept.-Dec. W--H-.

††**Skeleton Fan-flower** *Scaevola depauperata* Goodeniaceae. Rigid semi-shrub to 60cm. Sept.-April ---H-.

Prickly Fan-flower *Scaevola spinescens* Goodeniaceae. Rigid shrub. 50-90cm. Summer --PH-.

Cup Velleia
Velleia connata
Goodeniaceae.

Annual. 30-90cm. Oct.-Dec. ---H-.

Bell-fruit Tree
Codonocarpus cotinifolius
Gyrostemonaceae.

Shrub or tree to 12m. Sept.-Feb. W--H-.

Wheel-fruit
Gyrostemon australasicus
Gyrostemonaceae.

Shrub to 1m. Responds to fire. Sept.-Feb. W--H-.

Golden Pennants
Glischrocaryon behrii
Haloragaceae.

Tufted perennial to 45cm. Spring WBPH-.

Toothed Raspwort
Haloragis odontocarpa
Haloragaceae.

Annual 15-45cm. Oct.-Feb. ---H-.

Red Water-milfoil
Myriophyllum verrucosum
Haloragaceae.

Aquatic perennial. Summer W--H-.

Austral Bugle *Ajuga australis*
Lamiaceae. Perennial. 15-50cm.
Sept.-Feb. W-PHM.

***Horehound** *Marrubium vulgare*
Lamiaceae. Troublesome perennial to
60cm. W-PHM.

Scarlet Mint-bush *Prostanthera aspalathoides* Lamiaceae. Shrub 15-45cm. Colour variable. Sept.-Dec. WB---.

Small-leaf Mint-bush *Prostanthera microphylla* Lamiaceae. Shrub. 15-40cm. Sept.-Jan. W-PH-.

Erect perennial. 15-30cm. Spring-Autumn W-PH-.

Grey Germander
Teucreum racemosum
Lamiaceae.

Slender Westringia
Westringia eremicola
Lamiaceae.

Shrub to 1m. Aug.-Oct. WB---.

Shrub. 30-50cm. Spring W-PH-.

Stiff Westringia
Westringia rigida
Lamiaceae.

Slender Dodder-laurel
Cassytha glabella
Lauraceae.

Twining parasite. Summer WB---.

Twining parasite. Sept.-Dec. WBPH-.

Coarse Dodder-laurel
Cassytha melantha
Lauraceae.

Downy Dodder-laurel
Cassytha pubescens
Lauraceae.

Twining parasite. Oct.-April WB---.

Native Flax
Linum marginale
Linaceae.

Slender perennial to 60cm. Aug.-Nov. --P--.

Flax-leaf Logania
Logania linifolia
Loganiaceae.

Shrub. 30-60cm. Spring WB---.

Bare Logania
Logania nuda
Loganiaceae.

Leafless semi-shrub to 40cm. Spring W--H-.

†**Buloke Mistletoe** *Amyema linophyllum*
Loranthaceae. Parasitic on large casuarinas. Summer W--HM.

Box Mistletoe *Amyema miquelii*
Loranthaceae. Parasitic on eucalypts. Dec.-April W--H-.

Fleshy Mistletoe *Amyema miraculosum*
Loranthaceae. Parasitic on a variety of hosts. Sept.-April W--HM.

Drooping Mistletoe *Amyema pendulum*
Loranthaceae. Parasitic on eucalypts. Mar.-Oct. ---H-.

Wire-leaf Mistletoe
Amyema preissii
Loranthaceae.

Parasitic on acacias. Nov.-Mar. ---H-.

Harlequin Mistletoe
Lysiana exocarpi
Loranthaceae.

Parastic. Many hosts. Nov.-July W--HM.

†Desert Lantern
Abutilon otocarpum
Malvaceae.

Semi-shrub to 60cm. Sept.-May ---H-.

Chingma Lantern
Abutilon theophrasti
Malvaceae.

Erect annual to 1m. Sept.-May ---H-.

Australian Hollyhock
Lavatera plebeia
Malvaceae.

Annual or perennial to 1½m. July-Dec. W--H-.

Harrow Wattle
Acacia acanthoclada
Mimosaceae.

Rigid shrub to 1m. July-Sept. ---H-.

Gold-dust Wattle *Acacia acinacea*
Mimosaceae. Shrub. 30cm-1½m.
Aug.-Oct. W----.

Grey Mulga *Acacia brachybotrya*
Mimosaceae. Shrub to 2m. Sept.-Oct.
W-PH-.

Wallowa *Acacia calamifolia* Mimosaceae.
Shrub to 2½m. July-Sept. WBP--.

Hakea Wattle *Acacia hakeoides*
Mimosaceae. Shrub to 2m. Aug.-Sept.
WBPH-.

†††Wait-a-while
Acacia colletioides
Mimosaceae.

Rigid shrub to 2½m. Aug.-Oct. --PHM.

Small Cooba
Acacia ligulata
Mimosaceae.

Shrub 1-4m. Spring W-PH-.

†††Streaked Wattle
Acacia lineata
Mimosaceae.

Shrub to 1½m. Spring W----.

Shrub to 3m. Aug.-Oct. W--H-.

Manna Wattle
Acacia microcarpa
Mimosaceae.

Mallee Wattle
Acacia montana
Mimosaceae.

Shrub. 1-2m. Aug.-Oct. W-P--.

Shrub. 1-2m. Aug.-Oct. -B---.

Myrtle Wattle
Acacia myrtifolia
Mimosaceae.

Umbrella Wattle
Acacia osswaldii
Mimosaceae.

Shrub or tree to 5m. Nov.-Jan. W-PH-.

Golden Wattle
Acacia pycnantha
Mimosaceae.

Shrub or tree to 4m. Aug.-Oct. W----.

Nealie
Acacia rigens
Mimosaceae.

Shrub. 1-2m. July-Oct. WBPH-.

Bushy shrub to 1m. Aug.-Oct. W--H-.

Hard-leaf Wattle
Acacia sclerophylla
Mimosaceae.

Stiff shrub to 1m. Aug.-Oct. WB-H-.

Spiny Wattle
Acacia spinescens
Mimosaceae.

Tree to 10m. Near water. April-July ---H-.

Eumong
Acacia stenophylla
Mimosaceae.

†Three-nerved wattle
Acacia trineura
Mimosaceae.

Shrub. 2-5m. Aug.-Oct. W----.

Bramble Wattle
Acacia victoriae
Mimosaceae.

Shrub to 3m. Aug.-Nov. ---H-.

Spreading Emu-bush
Eremophila divaricata
Myoporaceae.

Erect shrub. 1-2m. Nov.-Feb. ---H-.

Common Emu-bush *Eremophila glabra*
Myoporaceae. Variable shrub to 1m. Aug.-Dec. W-PHM.

Berrigan *Eremophila longifolia*
Myoporaceae. Shrub or tree to 6m. All months W-PH-.

Sugarwood *Myoporum platycarpum*
Myoporaceae. Tree to 10m. Aug.-Dec. W-PHM.

Desert Baeckea *Baeckea crassifolia*
Myrtaceae. Erect shrub to 50cm. June-Sept. WBPH-.

Creeping Myoporum
Myoporum parvifolium
Myoporaceae.

Prostrate shrub to 30cm. Summer W--H-.

Broom Baeckea
Baeckea behrii
Myrtaceae.

Erect shrub to 2m. Sept.-Dec. WBPH-.

Mat Baeckea
Baeckea ericaea
Myrtaceae.

Spreading shrub to 20cm. Aug.-Nov. WB---.

†††**Prickly Bottlebrush** *Callistemon brachyandrus* Myrtaceae. Shrub or tree to 4m. Summer ---H-.

Snow Myrtle *Calytrix alpestris* Myrtaceae. Shrub to 1½m. Oct.-Nov. WB---.

Common Fringe-myrtle *Calytrix tetragona* Myrtaceae. Shrub. ½-1m. Oct.-Nov. WB-H-.

Red Mallee *Eucalyptus calycogona* Myrtaceae. Tree. 5-10m. June-Sept. W-P--.

Bull Mallee
Eucalyptus behriana
Myrtaceae.

Tree. 3-10m. Dec.-Jan. W---M.

Tree to 15m. Jan.-April -B---.

Brown Stringybark
Eucalyptus baxteri
Myrtaceae.

Dumosa Mallee
Eucalyptus dumosa
Myrtaceae.

Tree. 3-8m. Nov.-April WBPHM.

Tree. 2-8m. Jan.-Mar. WBPHM.

Slender-leaf Mallee
Eucalyptus foecunda
Myrtaceae.

White Mallee
Eucalyptus gracilis
Myrtaceae.

Tree. 3-10m. May-Sept. W-PHM.

Tree. 3-8m. Spring WBPH-.

Yellow Mallee
Eucalyptus incrassata
Myrtaceae.

Black Box
Eucalyptus largiflorens
Myrtaceae.

Tree. 10-15m. Near watercourses. Jan.-April W--H-.

Tree. 10-30m. June-Dec. WB---.

Yellow Gum
Eucalyptus leucoxylon
Myrtaceae.

Grey Box
Eucalyptus microcarpa
Myrtaceae.

Tree. 10-25m. Feb.-July W----.

Blue Mallee *Eucalyptus polybractea*
Myrtaceae. Tree. 3-9m. April-Sept.
--PH-.

Grey Mallee *Eucalyptus socialis*
Myrtaceae. Tree. 2-9m. Summer
WBPHM.

Muntries *Kunzea pomifera* Myrtaceae.
Prostrate shrub. Oct.-Dec. WB---.

Green Tea-tree *Leptospermum coreaceum*
Myrtaceae. Shrub. l-2m. July-Oct.
WBPH-.

Heath Tea-tree *Leptospermum myrsinoides* Myrtaceae. Shrub. 1-2m Sept. Nov. WB---.

Creamy Honey-myrtle *Melaleuca acuminata* Myrtaceae. Shrub. 1-2m. Spring WB-H-.

Salt Paperbark *Melaleuca halmaturorum* Myrtaceae. Shrub or tree to 8m. Salt lakes. Oct.-Dec. --PH-.

Moonah *Melaleuca lanceolata* Myrtaceae. Shrub or tree to 8m. Summer WBPH-.

Mallee Honey-myrtle
Melaleuca neglecta
Myrtaceae.

Shrub to 2½m. Nov.-Dec. WB---.

Broom Honey-myrtle
Melaleuca uncinata
Myrtaceae.

Slender shrub. 1-2m. Oct.-Dec. WBP--

Violet Honey-myrtle
Melaleuca wilsonii
Myrtaceae.

Shrub to 3m. Oct.-Dec. -B---.

Heath Myrtle
Micromyrtis ciliata
Myrtaceae.

Shrub to 40cm. Sept.-Dec. WPB--.

†Desert Jasmine
Jasminum didymum
Oleaceae.

Shrub. ½-2m. Summer ---H-.

Clove-strip
Ludwigia peploides
Onagraceae.

Creeping or floating perennial. Oct.-Dec. ---H-.

***Lesser Broomrape** *Orobanche minor*
Orobanchaceae. Erect leafless perennial. 15-30cm. Oct.-Nov. W--H-.

***Bristle Poppy** *Papaver aculeatum*
Papaveraceae. Annual. 15-60cm. Sept.-Dec. W--H-.

***Opium Poppy** *Papaver somniferum*
Papaveraceae. Annual. 30-90cm. Summer W--H-.

Sweet Appleberry *Billardiera cymosa*
Pittosporaceae. Climbing shrub. Oct.-Nov. WBPH-.

Sweet Bursaria
Bursaria spinosa
Pittosporaceae.

Shrub. 1-2m. Oct.-Dec. WBPH-.

Shrub or tree to 7m. Aug.-Nov. W-PHM.

Weeping Pittosporum
Pittosporum phyllyreoides
Pittosporaceae.

†Broom Milkwort
Comesperma scoparium
Polygalaceae.

Shrub to 60cm. Aug.-Nov. ---H-.

Blue-spike Milkwort
Comesperma calymega
Polygalaceae.

Perennial. 15-25cm. Nov.-Jan. W----.

Tangled Lignum
Muehlenbeckia cunninghamii
Polygonaceae.

Spreading shrub. 1-2m. Swamps. Aug.-Dec. W--H-.

Twiggy Lignum
Muehlenbeckia diclina
Polygonaceae.

Erect shrub to 1m. Sept.-April W--H-.

Pink Purslane
Calandrinia calyptrata
Portulacaceae.

Procumbent annual. Sept.-Oct. WB-H-.

†††Twining Purslane
Calandrinia volubilis
Portulacaceae.

Tangled annual. Sept.-Dec. -----.

Gland Flower
Adenanthos terminalis
Proteaceae.

Shrub to 90cm. Sept.-May WB---.

Silver Banksia *Banksia marginata*
Proteaceae. Shrub or tree. 1-4m. Spring-Autumn. WB---.

Desert Banksia *Banksia ornata*
Proteaceae. Bushy shrub. 1-2m. Mar.-Aug. WB---.

Slender Smoke-bush *Conospermum patens*
Proteaceae. Slender shrub to 90cm. Oct.-Dec. WB---.

Comb Grevillea *Grevillia huegelii*
Proteaceae. Rigid shrub. 1-2m. Sept.-Jan. W--H-.

Holly Grevillea *Grevillea ilicifolia*
Proteaceae. Shrub. ½-2m. Sept.-Nov.
WBPH-.

Desert Grevillea *Grevillea pterosperma*
Proteaceae. Shrub. 2-3m. Sept.-Jan.
WB-H-.

Lavender Grevillea *Grevillea lavandulacea*
Proteaceae. Shrub 45-90cm. Aug.-Oct.
WB---.

Desert Hakea *Hakea muelleriana*
Proteaceae. Shrub to 3m. Sept.-Nov.
WB-H-.

Shrub to 4m. Sept.-Jan. --PHM.

Silver Needlewood
Hakea leucoptera
Proteaceae.

Hooded Needlewood
Hakea tethrosperma
Proteaceae.

Shrub or tree to 10m. Spring W--H-.

Erect shrub to 3m. Spring -B---.

Furze Hakea
Hakea ulicina
Proteaceae.

Prickly Geebung
Persoonia juniperina
Proteaceae.

Shrub. 30-90cm. Summer -B---.

Horny Conebush
Isopogon ceratophyllus
Proteaceae.

Compact shrub to 50cm. Spring -B---.

Small-leaf Clematis
Clematis microphylla
Ranunculacea.

Perennial climber. Aug.-Oct. WBPH-.

Spreading shrub. 10-40cm. Spring WB-H-.

White Cryptandra
Cryptandra leucophracta
Rhamnaceae.

Silky Cryptandra
Cryptandra propinqua
Rhamnaceae.

Stiff shrub to 1¼m. July-Oct. ---H-.

Stiff shrub. 30-90cm. Winter WB-H-.

Prickly Cryptandra
Cryptandra tomentosa
Rhamnaceae.

Heath Spyridium
Spyridium eriocephalum
Rhamnaceae.

Spreading shrub. 40-90cm. Spring WB-H-.

Velvet Spyridium
Spyridium subochreatum
Rhamnaceae.

Shrub. ½-1m. Aug.-Nov. WBPH-.

Common Woodruft
Asperula conferta
Rubiaceae.

A tufted perennial to 15cm. Sept.-Dec. WB-H-.

Spindly shrub to 40cm. Spring WB-H-.

Blue Boronia
Boronia caerulescens
Rutaceae.

Common Correa
Correa reflexa
Rutaceae.

Shrub. ½-1m. Feb.-Aug. WB---.

Shrub to 50cm. Aug.-Oct. W----.

†††Narrow-leaf Wax-flower
Eriostemon angustifolius
Rutaceae.

Prickly Wax-flower *Eriostemon pungens* Rutaceae. Procumbent shrub to 30cm. Sept.-Oct. WB---.

††**Wilga** *Geijera parviflora* Rutaceae. Tree to 10m. April-Sept. ----M.

Desert Phebalium *Pheballium bullatum* Rutaceae. Slender shrub to 90cm. Sept.-Nov. WBPH-.

Broom Ballart *Exocarpos sparteus* Santalaceae. Shrub or tree to 4m. All months WB-H-.

Shrub or Tree to 4m. July-Dec. W-PH-.

Leafless Ballart
Exocarpos aphyllus
Santalaceae.

†Sweet Quandong
Santalum acuminatum
Santalaceae.

Shrub or tree to 5m. Summer ---HM.

Fruit.

†Sweet Quandong
Santalum acuminatum
Santalaceae.

Common Sour-bush
Choretrum glomeratum
Santalaceae.

A straggling shrub. 1-2m. Jan.-April -----.

A shrub 2-4m. Summer-Autumn -----.

††Northern Sandalwood
Santalum lanceolatum
Santalaceae.

††Northern Sandalwood
Santalum lanceolatum
Santalaceae.

Fruit.

Tree. 4-6m. Spring-Summer WBPH-.

Bitter Quandong
Santalum murrayanum
Santalaceae.

Bitter Quandong
Santalum murrayanum
Santalaceae.

Fruit.

Erect shrub. 1-2m. June-Nov. W----.

Wedge-leaf Hop-bush
Dodonaea viscosa
ssp. cúneata
Sapindaceae.

Slender Hop-bush
Dodonaea viscosa
ssp. angustissima
Sapindaceae.

Bushy shrub to 4m. Fruit W-PHM.

Cattle-bush
Heterodendrum
oleifolium
Sapindaceae.

Tree. 3-6m. Nov.-Jan. W-PHM.

Cattle-bush
Heterodendrum
oleifolium
Sapindaceae.

Fruit.

Purple Eyebright
Euphrasia collina
Scrophulariaceae.

Perennial. 20-30cm. Sept.-Dec. WB---.

Small Mud-mat
Glossostigma elatinoides
Scrophulariaceae.

Creeping semi-aquatic. Oct.-Feb. W--H-.

Austral Mudwort
Limosella australis
Scrophulariaceae.

Creeping semi-aquatic. Summer W--H-.

Blue Rod *Morgania floribunda*
Scrophulariaceae. Erect perennial. 30-90cm. All months. W-HM.

Small-leaf Ray-flower *Cyphanthera myosotidea* Solanaceae. Shrub. 30-40cm. Oct.-Nov. W-PH-.

Australian Box-thorn *Lycium australe*
Solanaceae. Spiny shrub. 60-90cm. All months ---H-.

***Tree Tobacco** *Nicotiana glauca*
Solanaceae. Shrub. 2-4m. All months W-PHM.

Velvet Tobacco *Nicotiana velutina* Solanaceae. Perennial to 90cm. All months W-PH-.

Oondoroo *Solanum simile* Solanaceae. Shrub. 1-2m. All months W--H-.

Creamy Stackhousia *Stackhousia monogyna* Stackhousiaceae. Perennial. 15-30cm. Aug.-Nov. WBPH-.

Pink Velvet *Lasiopetalum behrii* Sterculiaceae. Shrub. ½-1½m. Oct.-Dec. WB-H.

Slender Velvet-bush
Lasiopetalum baueri
Sterculiaceae.

Shrub to 1m. Spring WB-H-.

Annual. 2-7cm. Spring WB---.

Hairy Stylewort
Levenhookia dubia
Stylidiaceae.

Grass Trigger-plant
Stylidium graminifolium
Stylidiaceae.

Perennial. 15-90cm. Sept.-Jan. -B---.

Spreading shrub. 1-2m. All months. W-P--.

Mallee Rice-flower
Pimelea microcephala
Thymeleaceae.

Woolly Rice-flower
Pimelea octophylla
Thymeleaceae.

Erect shrub. 30-60cm. Spring-Summer WB---.

Erect shrub. 30-60cm. Sept.-Dec. -B---.

Heath Rice-flower
Pimelea phylicoides
Thymeleaceae.

Gaunt Rice-flower
Pimelea stricta
Thymeleaceae.

Slender shrub. 60-90cm. Aug.-Nov. WB---.

Shrub Violet
Hybanthus floribundus
Violaceae.

Shrub to 50cm. Aug.-Oct. WBPH-.

Nitre-bush
Nitraria billardiera
Zygophyllaceae.

Spreading shrub. 1-1½m. Saline soils. Spring --PHM.

Caltrop *Tribulus terrestris*
Zygophyllaceae. Prostrate annual.
Summer-Autumn W--H-.

Pointed Twin-leaf *Zygophyllum apiculatum*
Zygophyllaceae. Shrubby perennial.
20-40cm. Aug.-Oct. W-PHM.

Shrubby Twin-leaf *Zygophyllum aurantiacum* Zygophyllaceae. Shrubby perennial to 1m. Aug.-Dec. --PH-.

Pale Twin-leaf *Zygophyllum glaucum*
Zygophyllaceae. Annual to 20cm.
July-Dec. W-PH-.

Index

Index — continued

Index — continued

Index — continued

Index — continued

* = alien species.

† = vulnerable.
†† = endangered.
††† = rare.
(Captions only.)

Books with colour illustrations of Mallee plants:

Plants of Western New South Wales: G. M. Cunningham, W. E. Mulham, P. L. Milthorpe and J. H. Leight (N.S.W. Government) Printer, 1981).

Native Trees and Shrubs of South-eastern Australia: Leon Costermans (Rigby, 1981).

Flowers and Plants of Victoria: C. R. Cochrane, B. A. Fuhrer, E. R. Rotherham and J. H. Willis (A. H. and W. A. Reed).

Flowers and Plants of New South Wales and Southern Queensland: E. R. Rotherham, B. G. Briggs, D. F. Blaxell and R. C. Carolin (A. H. and W. A. Reed).

A Guide to Plants in Little Desert and Mt. Arapiles Area: F. J. C. Rogers (F. J. C. Rogers).

The Land Conservation Council's "Report on the Mallee Area Review", (Victoria Government Printer, 1987) is essential reading for anyone interested in the ecology of the Mallee.